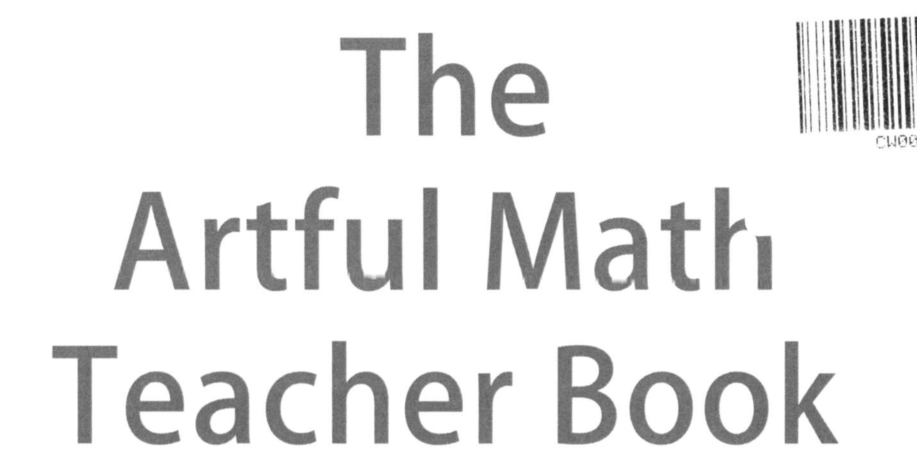

The Artful Math Teacher Book

Clarissa Grandi

About The Author

Clarissa Grandi is a mathematics teacher and geometric artist based in the UK. She is passionate about engaging learners in their mathematics through the exploration of pattern-making and mathematical art. She is the founder of ArtfulMaths.com, an online platform sharing ideas and resources to support the teaching of mathematical art and origami in schools. She can also be found on Twitter as @c0mplexnumber, where she posts and retweets creative mathematical images and ideas, and on Instagram as @clarissagrandi.art.

Introduction

The National Curriculum in England states that a high-quality mathematics education provides:

> ... an appreciation of the beauty and power of mathematics, and a sense of enjoyment and curiosity about the subject (DfE, 2013).

The lessons in this guide are designed to tick all those boxes. They have been developed with the intention of introducing learners to aspects of this multifaceted subject that often remain hidden: the beauty of geometric patterns; their links to the world around us; and the interesting- often quirky- mathematics that lies behind them.

Art and mathematics are usually thought to be at opposite ends of the subject spectrum. However, at the heart of both disciplines is the study of pattern, proportion and harmony. Perhaps then, we should envisage that spectrum as circular: so that, in fact, art and mathematics lie immediately alongside each other, at times overlapping. It is from this area of overlap that the beauty of mathematical art derives.

The wonderful thing about mathematical art is that the most beautiful geometric patterns can be produced without needing to be able to draw, or be 'good at art'. Mathematical art is accessible to learners of all ages: its algorithmic nature means that it simply requires the ability to follow instructions carefully and to use a pencil and ruler accurately. It is engaging, enriching, thoroughly enjoyable and is a great leveller in the classroom. Learners who may not normally shine in your lessons will take your breath away with their creativity. Those who struggle with their mathematics will experience the joy of success through their mathematical art-making.

The six Artful Math lessons that follow contain hands-on tasks that will develop important skills such as hand-eye co-ordination, manual dexterity and design thinking, which is a valuable form of problem-solving. Decisions need to be made about placement, size and color, all of which entail thinking about measurements, proportions and symmetry. Opportunities are also provided to draw out the mathematics underlying the patterns and to practice key content in the school curriculum.

Give learners time to play and explore, and to practice and develop their skills with their geometric tools. They deserve to feel confident about their geometry and to gain deeper understanding of its vital importance in the construction of our world, and of its intrinsic beauty.

Contents

ISBN 978-1-91109-318-3

How to Use this Guide

These six Artful Math lessons are designed for learners aged 9 to 16+. They are suitable for either single or double lessons (around 50 to 100 minutes of learning). For a short lesson you may simply decide to focus on the pattern making activity; but if you have more time, there is scope to draw out much more of the underlying mathematics.

Each lesson is accompanied by a downloadable PowerPoint presentation and templates available from Tarquin - to access these templates see details on the inside back cover of this book. Learners work on squared or plain paper, or on specially designed printable templates, also available to download. Alternatively, learners can work through their own individual copy of the Artful Math Activity Book that has been designed to accompany this publication.

The teacher notes should be read in conjunction with the relevant lesson PowerPoint. The notes are divided into the following sections:

Curriculum links An at-a-glance list detailing where each Artful Math lesson links to the content of a generic school mathematics curriculum. This will enable you to use the lessons to consolidate specific knowledge, skills and processes and to embed the lessons into your departmental scheme of learning. **Curriculum links in bold type are those that pertain to the higher-level subject knowledge required for optional extension tasks.**

Learning objectives A summary of the expected outcomes for learners. **Again, objectives given in bold type relate to the optional extension tasks.**

Resources A checklist of the resources required for each lesson.

Description of activity A summary of the lesson including any prior knowledge requirements for learners.

Suggested approach Step-by-step instructions for delivering the lesson, including key questions and prompts. Please note that these are for guidance only. The lesson PowerPoints are fully editable so that you can tailor the content of each lesson to the specific needs of your learners. As a result, no timings are given for the individual sections of the lesson. It is left to the teacher to decide how much time to devote to drawing out the related mathematics, and whether to include the investigations or extension tasks. These decisions will depend on the age and prior knowledge of your learners, and on the length of your lessons.

Support and extension Suggestions for how to support younger learners or those with additional needs to access the lesson, and suggested extension tasks for early finishers or older learners who may relish further challenge.

Further ideas Suggestions for extended projects that lead on from the content in the lesson, including cross-curricular projects, collaborative activities and opportunities for further research.

1 Curves of Pursuit

Curriculum links

Compare and classify 2D shapes based on their properties

Distinguish between regular and irregular polygons Draw

line segments accurately with a straight edge.

Measure accurately in centimetres and millimetres

Use straight edge and compass constructions

Draw shapes and nets accurately

Use geometric reasoning to solve problems

Use trigonometry to find angles in right-angled triangles

Learning objectives

- To construct a simple curve of pursuit inside a regular polygon (equilateral triangle, square, pentagon and/or hexagon)
- To subdivide a starting polygon into sub-polygons in order to construct a more complex curve of pursuit pattern within
- **To construct starting polygons with straight edge and compass**
- **To apply trigonometric ratios to find angles of turn in a square-based curve of pursuit**

Resources

- The Curves of Pursuit presentation
- 30 copies of the shoe-prints templates printed on card and cut out (optional)
- The printable support templates (or the accompanying Artful Math Activity Book)
- Pencils and straight edges
- Coloured pencils or pens (optional)
- Pairs of compasses (optional)

Description of activity

This lesson introduces curves of pursuit via the 'three predatory bugs' problem (also known as the mice problem). Learners are then shown some 2D and 3D examples for inspiration, before being led through a series of step-by-step instructions for drawing their own such curves. There is much space for individual creativity, as well as scope to draw out a good amount of mathematical thinking at different levels. Learners are provided with opportunities to practice visualisation, as well as to identify and classify 2-D and 3-D shapes. They will also gain useful practice in accurate drawing and measuring. Additionally, the lesson can be used to practice the standard straight edge and compass constructions; and, as an extension, a geometric reasoning challenge can be posed.

Prior knowledge requirement: Learners do not need to have met straight edge and compass constructions previously as there are 'quick start' templates provided. Knowledge of trigonometry is required to prove the geometric reasoning challenge, but learners may reason it through in other ways.

Suggested approach

1	Begin the lesson by giving learners the 'three predatory bugs' problem (slide 1). Ask them to predict and sketch the paths of the three bugs, encouraging learners to make multiple predictions if they wish.	The predatory bugs problem.... *Three predatory bugs are initially sitting at the corners of an equilateral triangle. All at once, each of the bugs begin crawling with equal speed directly toward the bug on their right.* *What is the path of each bug?* Demonstration of the pursuit paths of predatory bugs:
2	Invite learners to share their predictions and discuss them. At this stage you may have the time (and space) to use the shoe-print templates provided. If so, give three learners ten pairs each, stand them 3 to 4 metres apart at the vertices of an imaginary triangle, and ask them to perform the 'pursuit', laying a shoe-print on the floor after each step they take as they 'pursue' the student to their right. The curve of pursuit should emerge on the floor below them. Alternatively, use the hyperlink on slide 1 to see the pursuit in action. Go on to show the animations (slides 2 and 3), prompting learners to name the polygons in each case. These curves are called 'whirls'.	The curves made by the bugs' tracks are called 'whirls'. If we plotted the bugs' 'sightlines' after each step, what might the results look like?
3	Next ask learners to visualise the effect of plotting the bugs' 'sight-lines' at regular intervals: e.g. after every step. Show the animation on slide 4: this is how we will be constructing our curves of pursuit (the instructions follow on later slides).	
4	Now show learners the examples of 2D and 3D curves of pursuit (slides 5–16) drawing attention to the following: • Slide 6: this is the simplest pursuit curve – a single predator chasing its prey; perhaps a fox and a rabbit. Will the rabbit make it to its rabbit hole in time? • Slide 11: How have these two images been constructed? What is the same? What is different? The large triangles have been subdivided into 9 smaller triangles and the curves of pursuit then constructed in the smaller triangles. In the left-hand image, the direction of turn has been alternated, leading to a 'scallop' effect; whereas in the right-hand image the direction of turn remains the same throughout (counterclockwise) resulting in a 'twist' effect. • Slide 12: How has this image been constructed? A hexagon subdivided into six smaller triangles. Has direction of turn been alternated or not (scallop or twist effect)? • Slides 13–16: Some examples of 3D curves of pursuit. What do learners notice? What do they wonder?	

5	Now demonstrate the step-by-step instructions for drawing a curve of pursuit within a square (slides 17–23). Stress the need for learners to draw their marks working in the same direction each time (clockwise or counter-clockwise) and for the new set of marks to be made on the new smaller, inner polygon each time. 1cm or 0.5cm steps are easiest to start with.	
6	Learners are now ready to construct and color their curves. You may wish to propose that they start with the simpler polygons: squares or triangles, before progressing onto pentagons, hexagons or experimenting with sub-divided polygons. Depending on learners' prior knowledge you can ask them to construct their starting polygons using compass and straight edge, or alternatively just use the 'quick-start' templates (see below).	
7	On completion you can set learners the geometric reasoning challenge on slide 27 (see the notes below).	

Support and extension

To eliminate the need for initial compass and straight edge constructions, printable templates are provided for additional support, or for a 'quick start'. These templates provide the starting vertices for drawing equilateral triangles, squares, pentagons and hexagons.

Learners can investigate the sequence of decreasing right-angled triangles that occur within a square-based curve of pursuit (slide 27). What happens to the size of the angle of turn the 'predators' make after each step as the sequence progresses and they approach the centre? Can they prove their conjecture? This provides a useful opportunity to discuss the difference between a demonstration (measuring the angles with a protractor) and a proof (using trigonometry).

Further ideas

Tessellation: If learners use starting polygons with the same side length, these can be tessellated to produce a stunning classroom display. As an extension, this activity could be used to draw out understanding about the link between the internal regular polygon angles and resulting tessellation possibilities. Groups could be challenged to collaboratively plan a semi-regular tessellation display before creating the pursuit curves for the display.

Nets: Print out blank nets of the Platonic solids onto card, and invite learners to decorate them with curves of pursuit before assembling them. Hang them from the ceiling to make Platonic solid mobiles. To add a greater degree of challenge, learners could be asked to construct their own nets with compass and straightedge, before decorating them with pursuit curves.

2 Impossible Objects

Curriculum links

Compare and classify 2D and 3D shapes based on their properties

Draw line segments accurately with a straight edge

Measure accurately in centimetres and millimetres

Use ruler and compass constructions

Draw 3D shapes on an isometric grid

Learning objectives

- To use straight edge compass to construct an 'impossible' equilateral triangle: the Penrose triangle
- To extend this technique to constructing an 'impossible rectangle'
- To extend this technique to constructing a 'blivet'
- **To draw impossible objects on an isometric grid**

Resources

- The Impossible Objects presentation
- The printable instructions handout (or the accompanying Artful Math Activity Book)
- Pencils and straight edges
- Pairs of compasses
- Plain paper and isometric dotty paper (or the accompanying Artful Math Activity Book)
- Colored pencils or pens (optional)
- The means of showing a YouTube video (optional)

Description of activity

This lesson introduces the concept of impossible objects starting with the simplest example, the Penrose triangle. Learners are shown several examples for inspiration, before being led through a series of step-by-step instructions for constructing their own such objects. There is much space for student creativity, as well as scope to draw out a good amount of mathematical thinking at many different levels. Learners are provided with opportunities for discussion, as well as to identify and classify 2-D shapes and use geometric terminology. They will also gain useful practice in accurate drawing and measuring and consolidating some of the standard straight edge and compass constructions.

Prior knowledge requirement: If you wish learners to construct their impossible objects with straight edge and compass they should be familiar with the construction of the equilateral triangle. If learners have not yet met straight edge and compass constructions they can instead draw the shapes on isometric paper.

Suggested approach

1	Begin the lesson by showing learners the Penrose triangle (slide 2). Explain that this is an example of an 'impossible object': it can be drawn but it can't be made. Ask if they can explain why that might be. Some learners will be difficult to convince! Invite explanations from their peers, encouraging the use of correct geometric terminology, including the terms 'equilateral triangle', 'vertices', 'edges'. The gif on slide 3 may support the discussion.	The Penrose Triangle
2	Further examples of Penrose triangles follow on slides 4 – 6. Draw attention to the importance of the different depths of shading that contribute to the 3D effect. Slide 7 shows a metal sculpture that demonstrates clearly the construction of the illusion.	Penrose Triangle dice illusion
3	Other objects can be impossible too... Show slides 9 and 10, an 'impossible rectangle' and 'impossible cuboid'. These are more obviously impossible. Again, invite explanations from the learners encouraging correct terminology. Why would these structures be impossible to construct in reality?	Impossible wooden cuboid
4	Slide 11 shows the 'blivet', a form of impossible trident. What do learners notice? How many prongs has it got? Slides 12 and 13 go on to show some creative variations on the basic blivet structure.	The Blivet
5	The next set of slides shows more impossible constructions including Penrose stairs and the classic waterfall illusion. Slide 16 shows a variation of M.C. Escher's famous 'Waterfall' lithograph. What do learners notice? Can they identify what makes the structure 'impossible'?	Impossible waterfall, based on a painting by MC Escher, 1961
6	You may have time to follow this discussion by showing the construction of a real-life Escher's Waterfall illusion via the YouTube links on slide 17 (six minutes). Alternatively, you may wish to save these for the end of the lesson.	Real-life Escher's waterfall: http://www.youtube.com/watch?v=0v2xnl6LwJE Real-life Escher's waterfall explanation 1: http://www.youtube.com/watch?v=jpclla2hKRo&NR=1&feature=fvwp Real-life Escher's waterfall explanation 2: http://www.youtube.com/watch?v=Z6OeQtnultc

7	Now demonstrate the step-by-step instructions for drawing the Penrose triangle (slides 19–23). (If your learners have not yet met straight edge and compass constructions, then you can instead use isometric paper – section 9 below.) Ensure pens are down while you go through the instructions for the first time. Then, either return to slide 19 and walk the learners through the process again as they draw their triangles, or distribute the printable instructions hand-out (also included in the accompanying Activity Book).	The Penrose triangle 11cm Construct an equilateral triangle with side length 11cm.
8	Do the same for the impossible rectangle and blivet, slides 24–32. Encourage learners to shade their objects to create a 3D effect. Challenge learners to be creative with their blivets – what can they turn them into?	The impossible rectangle 7.5cm 7.5cm 15cm Construct two identical equilateral triangles on a 15cm base (each will have side length 7.5cm).
9	The final two slides show how to draw impossible objects on isometric paper. After showing learners the 'construction' for the Penrose triangle, perhaps challenge them to draw the impossible rectangle unaided, and then to draw other shapes such as the impossible cuboid, or shapes of their own devising.	Challenge: now try the impossible rectangle What other impossible constructions can you draw on your isometric grid?

Support and extension

To eliminate the need for initial compass and straight edge constructions, younger learners can draw their impossible objects on 'triangle dotty' isometric paper (see the final section of the presentation). To assist with this, you can draw attention to the three 'V' shapes that make up the Penrose triangle, and the two 'U' shapes that make up the Impossible Rectangle. However, learners may come up with their own ways of visualising and constructing the shapes.

Alternatively, you can use the isometric grid task as an extension for those who have finished constructing their impossible objects with straight edge and compass. Are they able to draw the shapes on isometric paper without first seeing the instructions?

Further ideas

Challenge learners to create optical illusions by arranging multilink cubes or dice and photographing them from an angle that makes them look as if they create a 3D Penrose triangle. Remind them of the structure shown in the gif on slide 3 and the sculpture on slide 7. Some learners may be adept at using photo editing software to enhance the illusion further.

3 Perfect Proportions

Curriculum links

Use Pythagoras' Theorem to find lengths in right-angled triangles

Use surds and their decimal approximations

Compare lengths using ratio notation and scale factors

Express the division or multiplication of two quantities as a ratio or fraction

Form and solve quadratic equations using the quadratic formula

Recognise and use the Fibonacci sequence

Use straight edge and compass constructions (circular arcs)

Calculate angles of sectors of circles

Learning objectives

- **To use Pythagoras' Theorem to find the Golden Ratio in the Golden Rectangle**
- **To use ratios to form and solve a quadratic equation to prove the Golden Ratio**
- To recognise and use the Fibonacci sequence to find the Golden Ratio
- To construct a Golden Spiral
- **To calculate the Golden Angle using arc lengths**
- To understand the link between phyllotaxis and the Golden Angle in plant growth

Resources

- The Perfect Proportions presentation
- Squared paper for Investigation 1
- Pencils, straight edges and pairs of compasses
- The Golden Spiral template (or the accompanying Artful Math Activity Book)
- Colored pencils or pens (optional)

Description of activity

This series of tasks leads learners through various investigations that reveal the occurrence of the Golden Ratio (Phi) in the world around us: in geometrical proportions, in the Fibonacci sequence, in art and architecture, and in the natural world. There is enough content for two to three lessons, depending on what you wish to include: enough, perhaps, to fill a whole Phi day!

Learners will also construct a Golden (or Fibonacci) Spiral, and, if time, combine several such spirals in a design of their own devising.

Prior knowledge requirement: Two of the investigations involve the use of Pythagoras' Theorem, one involves the forming and solving of a quadratic equation, and one involves some idea of using arc lengths to calculate subtended angles in a circle. However, the presentation can easily be amended to omit certain investigations depending on the age and prior knowledge of the learners. Younger learners can focus on finding the ratio in the Fibonacci sequence and drawing the resulting spiral.

Suggested approach

1	Begin the lesson with the quick poll on slide 2. After the vote, reveal the rectangle that is said to be the most perfectly proportioned: the Golden Rectangle. Was it the favourite? Explain that this rectangle is proportioned according to the Golden Ratio, Phi (φ): if the shortest side has length 1, the longer side has length Phi. Ask what it might mean for something to be perfectly proportioned. You may wish to show the quote by Plato on slide 3.	The perfect proportion? Which rectangle do you like best?
2	Investigation 1: Constructing the Golden Rectangle and finding the ratio using Pythagoras' Theorem, slides 4–13. Learners will need squared paper on which to draw their starting square, and pairs of compasses for the arc. Ask them to choose an even number of squares for the side length of the square so that its base midpoint is easily identified. Challenge learners to calculate the base length of the rectangle when the side is given a length of 1 unit. This is Phi. You may wish to discuss the irrationality of Phi at this point.	Can you find perfection? 1 If the shorter side is 1 unit, can you work out the exact length of the longer side?
3	Investigation 2: Using ratios to form and solve a quadratic equation to prove the Golden Ratio, slides 14–17. Slides 14 and 15 introduce the Golden Ratio as the ratio of two parts and their whole. Slide 16 then supports learners to set up the situation as an equation. A hint is provided if needed. Slide 17 gives the solution. If your learners have not yet met quadratic equations, this section can be omitted, or shown for interest only.	Can you prove perfection? Let the length of the shorter part be 1. Then the larger part must have length $1 \times \varphi = \varphi$... 1 φ $1 + \varphi$ What is the length of the whole? Can you form an equation to find φ? Remember $\times \varphi =$
4	The Golden Ratio in art and architecture, slides 18–24. Stress that there is some debate as to whether the Golden Ratio really does appear in certain buildings and paintings, but that many artists and architects are known to have experimented with using it in their work. Ask learners to consider where the Golden Rectangle might be hiding in each image, then reveal.	Phi in art and architecture • Artists and architects throughout the ages have experimented with the golden ratio • There are ongoing debates about which great monuments and artworks of the past *really do* have the golden ratio in their measurements • Here are a selection of those that some mathematicians argue *do* exhibit the golden ratio...

5	Investigation 3: Finding the Golden Ratio in the Great Pyramid of Giza, slides 25–27. Another opportunity for learners to apply Pythagoras' Theorem. Explain that the cubit is based on the length from the tip of the middle finger to the bottom of the elbow, and is approximately equivalent to half a metre. Once learners have found the length of the hypotenuse to the nearest cubit, ask them to consider where the ratio might be hiding: we know we multiply by Phi to create new side lengths, how could we 'undo' this to find Phi?	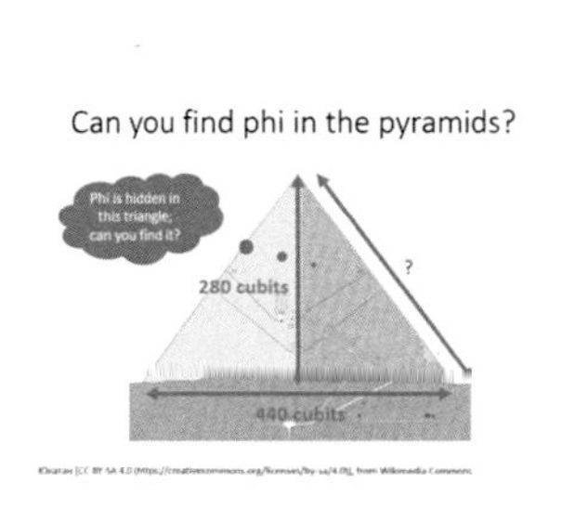
6	Investigation 4: Phi and Fibonacci, slides 28–30. Project the first few terms of the sequence and ask learners to explain the term-to-term rule and generate the next few terms. Next ask learners to investigate the ratios of consecutive terms – what do they notice? Phi again! Slide 31 shows the oscillating nature of the sequence of ratios as they tend toward Phi.	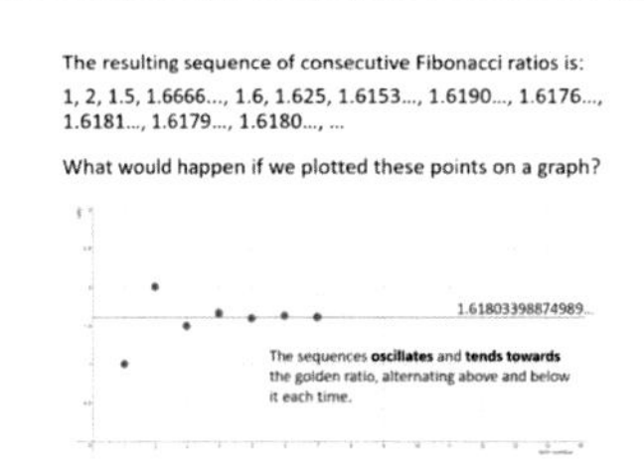
7	Drawing the Fibonacci Spiral, slides 31–33. Use the animation to demonstrate how arranging squares with Fibonacci side lengths and then drawing quarter circles through them creates the Golden Spiral. Hiding in the spiral are ever larger rectangles that tend towards the Golden Rectangle with its Phi-proportioned side lengths. Learners can now construct the spiral for themselves using the Golden Spiral template. Remind learners to check carefully where the compass point is placed for each quarter circle drawn. If you have time for them to get creative, hand out more squared paper and show the 'More Golden Spirals' slide for inspiration. Alternatively, early finishers can investigate Fibonacci 'backwards' on slide 34.	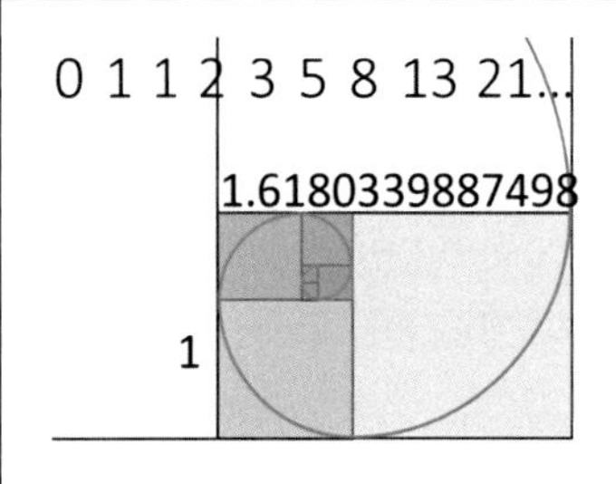 Fibonacci 'backwards' • Can you find the next terms of the sequence where each term is the sum of the two terms just ***after*** it? Start, as before, with 0, 1, ... 0, 1, -1, 2, -3, 5, -8, 13, -21... This is known as the **nega-Fibonacci** sequence, although some people call it the **Iccanobif** sequence...
8	Investigation 5: The Golden angle and Phi in nature, slides 35–45. Lead learners through the calculation for the Golden Angle using the circle with arc lengths in the ratio of 1 to Phi – or set it as a problem-solving challenge if they have the pre-requisite knowledge. Explain that this is a common angle for plants to grow new leaves, petals and seeds at, in order to maximise use of space and access to sunlight (a process called phyllotaxis). Show the gif on slide 38 which demonstrates the spirals that emerge from this process, followed by the images on slides 39-45. Invite learners to count the spirals on slides 40 and 41 – what do they notice? Consecutive Fibonacci numbers! Challenge them to find the Fibonacci numbers hidden in the spirals in the other images that follow. You may wish to use printed copies of the slides for this.	 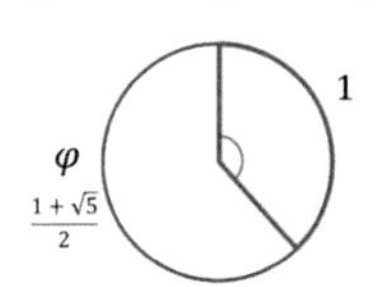 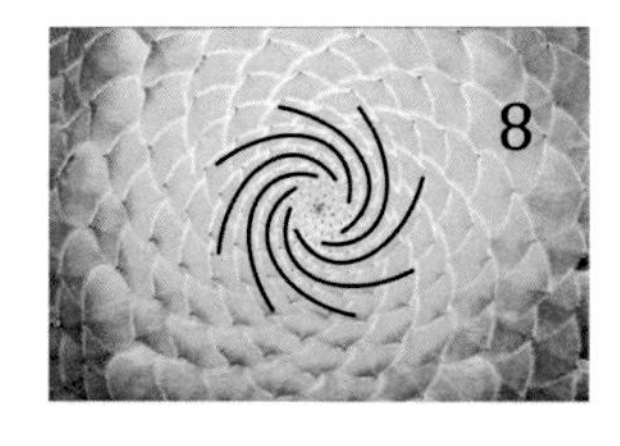 

9	You could finish this section by showing learners the fantastic 'Nature by Numbers' video by Cristóbal Vila – the YouTube link is given on slide 46. Please note that a link between the Nautilus shell and the Golden Spiral is suggested in the video, but not explained fully. The Golden Spiral is a logarithmic spiral with growth factor φ, whereas the Nautilus is a logarithmic spiral with growth factor $\ln(\varphi)/\pi$. Finally, slide 47 contains links to other interesting videos on Phi which could be used as part of a research project or homework.	Further exploration Vi hart on Spirals, Fibonacci and being a plant 1-3: • https://www.youtube.com/watch?v=ahXIMUkSXX0 • https://www.youtube.com/watch?v=lOIP_Z_-0Hs • https://www.youtube.com/watch?v=14-NdQwKz9w Numberphile on why the Golden Ratio is so irrational: • https://www.youtube.com/watch?v=sj8Sg8qnjOg

Support and extension

If learners are not yet proficient with compasses, the Golden Spiral can be drawn freehand on the template provided.

Further ideas

Paula Beardell Krieg (@PaulaKrieg on Twitter) has devised a wonderful project for playing with arrangements of Phi-scaled hexagons: https://bookzoompa.wordpress.com/2018/10/10/hexagons-and-the-golden-ratio. And no need to stop at hexagons! You could use the enlargement facility on a photocopier with a scale of 162% to create copies of other shapes scaled in the Golden Ratio for your learners to explore. What do they notice? What do they wonder?

4 Mazes and Labyrinths

Curriculum links

Use straight edge and compass constructions (circular arcs)

Explore and make conjectures about the generalisations that underlie patterns

Find the nth term of a linear sequence

Learning objectives

- To construct a branching maze
- To construct a labyrinth with straight edge and compass
- **To make and test conjectures about the sequence that underlies the construction of a labyrinth**
- **To deduce the nth term expression for the labyrinth construction sequence**

Resources

- The Mazes and Labyrinths presentation
- Squared paper and plain paper (or the accompanying Artful Math Activity Book)
- The printable instructions handouts (or the accompanying Artful Math Activity Book)
- Pencils, straight edges and erasers
- Colored pencils or pens (optional)
- Pairs of compasses for drawing the labyrinths

Description of activity

This lesson teaches how to construct branching mazes and traditional labyrinths. Learners are first introduced to mazes and labyrinths and the distinction between them, before being shown historic and creative examples of different types of each structure. They are next taken through a series of step-by-step instructions for drawing their own branching maze. Finally, there is the opportunity to learn to draw a traditional labyrinth with a pair of compasses. There is much space for creative problem solving in this lesson, as well as useful practice in accurate construction with compasses. As an extension there is an investigation into the mathematical sequence upon which labyrinths are built.

Depending on the length of your lessons, you may wish to leave drawing labyrinths for a separate lesson. It is a particularly good activity to do before introducing constructions, as it provides learners with useful practice with their compasses.

Prior knowledge requirement: For the extension challenge it would be useful for learners to have met the concept of the nth term, but it is not essential.

Suggested approach

1	Begin the lesson by introducing the objective and showing learners the two types of construction that they will be learning to draw (slide 1).	Drawing your own Mazes and Labyrinths
2	Explain the distinction between the two types of 'maze' (slide 2) and summarise their brief history.	Mazes and labyrinths According to one definition, **mazes** have many branching paths, with only one path leading to the centre or exit, whereas **labyrinths** are single-pathed (unicursal). Some of the earliest mazes and labyrinths we know of were found in Egypt and in Crete, dating back over 4000 years. One of the most famous of these is the seven-circuit Cretan labyrinth, which we shall be learning how to draw. In Greek mythology the Cretan king Minos owned a labyrinth in which lurked the Minotaur – a half man, half bull creature who ate anyone lost in the labyrinth... The Romans built many mosaic labyrinths, which were typically found in the entrance halls of their villas. Mazes are still very popular today, with many large ones found in the grounds of parks and stately homes.
3	Next go through the set of examples (slides 3–10) drawing learners' attention to their similarities and differences, their age, their size, and their construction materials and methods – such variety!	Mazes and labyrinths Thinker's brain as a maze of choices, by Bill Sanderson
4	**Drawing a branching maze:** Ensure pens are down while you go through the instructions for the first time. • Slides 13–15: Stress that ***heavily shaded*** squares will not be erased, but that ***very lightly shaded*** squares may be erased at a later stage, so learners should not press very hard with their pencils when shading these. • Slides 16–17: The solution path is then created by erasing some of the very lightly shaded squares. It may seem counter-intuitive that this is done first. However, once the solution path is in place, it is 'disguised' by the addition of alternative routes that lead to dead ends. • Slides 18–19: Learners complete their branching maze by heavily shading in the remaining unshaded 'wall' squares.	Drawing a branching maze Next, ***very lightly*** shade in the next set of alternate squares, but on every row this time. These squares will either be part of the paths, or part of the walls. Drawing a branching maze Time to create the path! Make an entrance by erasing one of the lightly shaded squares on the edge. Then continue erasing lightly shaded squares to make a winding path to an exit. Drawing a branching maze Next heavily shade in the remaining lightly shaded wall squares. Finally label the entrance and exit. Start Finish

<table>
<tr><td>5</td><td>Once you have explained the process in full, either return to slide 12 and walk the learners through the process again as they draw their mazes, or distribute the printable instructions hand-out for reference.</td><td>Drawing your branching maze</td></tr>
<tr><td colspan="3">Depending on timings, you may wish to leave drawing labyrinths for another lesson.</td></tr>
<tr><td>7</td><td>Drawing the seven-circuit labyrinth (slides 20–31): Again, ensure you go through the instructions once fully before allowing learners to begin. Then, walk learners through the process again alongside their construction, or distribute the second instructions hand-out.</td><td>Drawing your seven-circuit labyrinth</td></tr>
<tr><td>8</td><td>On completion you can set learners the challenge (slide 32) of scaling up their labyrinths (see the notes below).</td><td>Drawing a labyrinth
Can you work out how to scale your labyrinth up or down?
How many concentric paths will be in the next biggest version?
Can you construct it?
Can you predict the number concentric paths for the 5th labyrinth?
The 12th? The nth?</td></tr>
</table>

Support and extension

A handout with step-by-step instructions for both types of construction is provided for reference.

Squared paper with smaller squares (0.5 – 0.7 cm) allows for more intricate branching mazes, although 1cm square paper may be more appropriate for younger learners or if you have less time.

The Labyrinth sequence problem (slide 32) provides a useful opportunity for conjecture and generalisation. Ask learners to discuss the sequence for scaling labyrinths up or down. Can they describe what happens? How many concentric paths (circuits) will be in the next term? Can they construct it? Can they predict the number circuits for the 5th labyrinth? The 12th? The nth?

Further ideas

Challenge your class to construct a walkable labyrinth in the school grounds using chalk, or perhaps found objects such as stones or twigs. They will need a length of rope to use as an enormous pair of compasses for drawing the arcs.

5 Epicycloids

Curriculum links

Identify and use circle definitions and properties, including radius, diameter and circumference

Apply and use the multiplication tables

Draw line segments accurately with a straight edge

Recognise and use reflection and rotational symmetry

Identify multiples

Explore and make conjectures about the generalisations that underlie patterns

Learning objectives

- To recall and practice multiplication tables (2s, 3s, 4s and beyond)
- To understand the concept of modulo arithmetic
- To construct a cardioid, a nephroid and an epicycloid (and perhaps more)
- **To make and test conjectures about the generalisations that underlie the patterns**

Resources

- The Epicycloids presentation
- The printable 60-point circle Epicycloid templates (or the accompanying Artful Math Activity Book)
- Pencils and straight edges
- Colored pencils or pens (optional)

Description of activity

This lesson introduces cardioids, nephroids and epicycloids (group name epicycloids) via a visualisation problem that draws out thinking about circles and their properties. Examples of the cardioid and where it appears naturally in the world around us are shown. The concept of modulo arithmetic is explained and learners are then led through a series of step-by-step instructions for filling out their multiplication mapping tables. The construction of the cardioid is clearly demonstrated enabling learners to go on to construct and decorate their own such curves. Learners gain useful practice in recalling multiplication tables and in drawing straight lines carefully and accurately.

Thereafter there is scope to draw out further mathematical thinking through the setting of investigative prompts about the properties of the epicycloid patterns and their links with the multiples involved.

Prior knowledge requirement: Knowledge of basic circle terminology is useful for the introductory problem. The extension investigation requires understanding of the concept of multiples.

Suggested approach

1	Begin the lesson with the 'circle around a circle' problem on slide 2. Ask learners to predict and sketch the path of the point on the circumference of the blue circle when it rolls itself around the outside of the red circle. What if the outer blue circle was smaller? Bigger? What would the condition be for the point to touch the red circle exactly twice? Three times? At this stage you may have the time to use some circular manipulatives with which learners can test their predictions. Two pence pieces are good, or large counters. Alternatively, learners could draw and cut out their own paper circles, providing an opportunity to include practice with compasses.	Circle around a circle What would the condition be for the point to touch the red circle exactly twice? Three times? What is the path traced by the point on the circumference of the blue circle, when the blue circle is rolled around the outside of the red circle?
2	Invite learners to share and discuss their conjectures. Then show the animation on slide 3. Explain that this curve is called a Cardioid due its heart shape, and it is a member of a family of curves called epicycloids.	
3	Show the animations on slides 4–6. Explain that the points where the path touches the inner circle are called 'cusps'. Introduce the two-cusped Nephroid curve (named after its kidney shape) which is produced when the radius of the outer circle is half that of the inner, and the three-cusped Epicycloid of Cremona, produced when it is a third of the size. Also discuss the two curves produced from larger outer circles. Were learners' predictions correct?	
4	Now explain that the lesson will focus on drawing and exploring these epicycloids, starting with the Cardioid. This curve arises naturally in the world around us: for example, in the reflection of a point of light off a circular surface (slides 7 and 8), and in the Mandelbrot set fractal (slide 9).	You can sometimes spot a **cardioid** on the surface of a cup of coffee!
5	The cardioid is drawn by plotting the reflected rays that create it, using a 60-point circle and mapping the points to the two times table modulo 60. Slide 11 explains the concept of modulo arithmetic in this context.	This process of 'wrapping around' times tables and starting again is called 'modulo arithmetic'. We will 'map' $n \to 2n$ Mod 60 (the two times table mod 60).

6	Now take learners through the process for filling out their mapping tables and drawing their cardioids (slides 12–17). Stress the need to check the entries in their tables carefully before starting drawing, and to score out the pairs of points as they connect them.	Keep scoring out the matches as you go so that you don't lose your place. You'll be drawing a lot of lines!
7	If time, learners can explore what happens with the mappings $n \rightarrow 3n$ Mod 60 and $n \rightarrow 4n$ Mod 60 (slides 18–19). They will recognise the Nephroid and the Epicycloid of Cremona. Take the opportunity to draw attention to the symmetries of the patterns. How many axes of reflection are there? Where are they? And what are the orders of rotational symmetry? Encourage learners to color in their patterns in such a way as to highlight these inherent symmetries.	$n \rightarrow 3n$(mod 60) Nephroid; $n \rightarrow 4n$(mod 60) Epicycloid of Cremona

Support and extension

A set of investigative prompts are provided as an extension activity on slide 21. Learners can explore the different mappings using the interactive tool created by Jonathan Hall (@StudyMaths on Twitter) at https://mathsbot.com/tools/cardioids. The prompts draw out thinking about the multiples (and non-multiples) in the patterns. An answer sheet for the investigation is also provided.

Further ideas

Learners can use the flashlight facility on their smart phones to generate a cardioid inside a glass of water. What happens if they increase the number of light sources?

Learners can cut out their cardioids to decorate mathematical cards for Valentine's Day. A5 copies of the templates may be a more appropriate size for this.

6 Parabolic Curves

Curriculum links

Understand a locus as a set of points that obey a given rule Use straight edge and compass constructions (reasoning)

Recognise and use reflection and rotational symmetry

Draw line segments accurately with a straight edge

Use geometric reasoning to solve problems

Use algebraic notation and the Cartesian plane

Use Pythagoras' Theorem to find lengths in right-angled triangles

Learning objectives

- **To understand the parabola as a locus of points**
- To construct the parabola by folding paper
- To construct parabolic curves from straight lines
- **To prove the parabola is the graph of a quadratic relationship**

Resources

- The Parabolic Curves presentation
- Printouts of slide 26 (and slide 27 for the optional extension challenge)
- The parabolic axes templates (or the accompanying Artful Math Activity Book)
- Pencils and straight edges
- Colored pencils or pens (optional)

Description of activity

This lesson introduces the parabolic curve as the locus of points equidistant from a point and a line. Learners first draw an approximation of the curve based upon this principle, and then fold it from paper. Next, learners are shown where parabolic curves appear in the world around us, and why they are important. Then learners are guided through constructing these curves from a series of straight lines. Once the basic principle is mastered, there is the opportunity for creative problem solving where learners use the printable templates as a basis for their own parabolic designs. There is an optional extension challenge to prove the parabola is the graph of a quadratic function, suitable for learners at KS4 or KS5.

Prior knowledge requirement: Ideally learners will have met the concept of a locus beforehand, but this can also be explained by the teacher during the session. The optional extension challenge will require the application of Pythagoras' Theorem and familiarity with graphing on the Cartesian plane.

Suggested approach

1	Begin the lesson by outlining the objectives and introducing the parabola as the locus of all points equidistant from a given line and a point that is not on the line (slide 2). If learners have not met the concept of 'locus' before, simply explain it as a set of points that obey a rule; and, if necessary, explain the term 'equidistant'. Introduce the terms 'focus' and 'directrix'.	What is a parabola? **The locus of all points *equidistant* from a given line and a point that is not on the line** The point is called the focus The line is called the directrix
2	Project slide 3 and hand out copies of the printable version (slide 26). Task learners with placing points on the vertical dashed lines that they estimate to be the same distance from the focus as from the directrix. They may wish to use straight edges to help with this, or to just place points by eye. Once they have had a go, invite learners up to the board to draw their points on the projected slide (if feasible). Then reveal the animated points and parabola (slide 4) – how accurate were their estimations?	Place a cross on each vertical line at a point that looks to be the same distance from the blue **directrix** as from the red **focus**
3	Next, walk learners through the folded parabola task (slides 5–8) and encourage them to complete at least 5 more folds independently. After reminding them about the definition of a locus, ask them to think about why the creases create a parabolic curve. Draw out the idea that the creases are halfway lines (perpendicular bisectors) between a specific, chosen point on the directrix, and the focus. Thus, any point where two creases intersect must be the nearest equidistant point between the directrix and focus on both lines, and hence lie on the parabola. This concept is challenging, so take it as far as you think appropriate.	You can also fold a parabola from a rectangular piece of paper. Keep doing this from different points along the bottom edge. What happens? Why? Do the same from another point on the bottom edge. You can also fold a parabola from a rectangular piece of paper. Draw a dot about 3 to 4 cm above the bottom edge of your paper – this will be your **focus**
4	Now introduce learners to some of the other situations in which parabolic curves appear in the world around us, and why their unique shape leads to important applications (slides 9–13).	Why are parabolas important? • Parabolic arches are widely used in bridge design as they are able to support heavy loads over wide spans. • They are also used in cathedrals and other large buildings. The Tyne Bridge

<table>
<tr>
<td>5</td>
<td>Constructing the curves: explain that we can produce parabolic curves by drawing the straight lines as well as by folding them. And furthermore, that we can combine translations, reflections and rotations of the basic curves to produce beautiful designs. Show slide 14 and ask learners to identify the transformations involved in the designs. Where are the axes of reflection symmetry? And what are the orders of rotational symmetry? Then talk through the instructions for constructing the simplest version (slides 15–16)

Learners can now construct their own curves on the parabolic axes templates. Once learners are confident with the basic principle you can let them experiment with the more complex templates. Show slide 17 for inspiration, but challenge learners to come up with their own designs too. Encourage the use of color or shading to highlight the internal symmetries of their patterns.</td>
<td>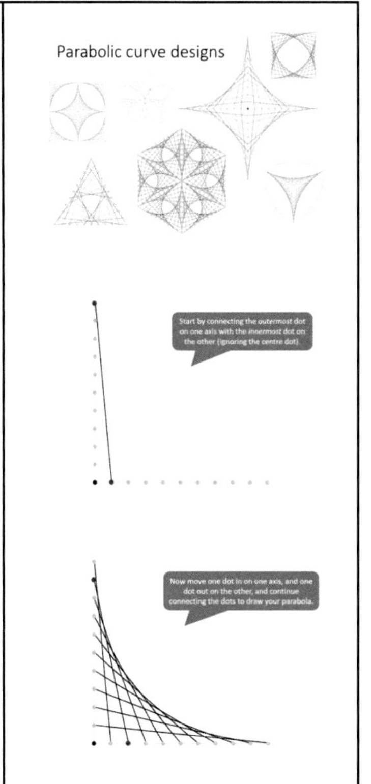
</td>
</tr>
<tr>
<td>6</td>
<td>On completion of their designs, some learners may be interested in the challenge of proving algebraically that the locus definition of a parabola produces a set of points obeying a quadratic relationship. The problem is outlined on slide 18 and there is a printable copy of the problem on slide 27. Prompts are given on slides 19 and 20 and the solution follows on slides 21–24. This challenge requires the application of Pythagoras' Theorem and familiarity with graphing on the Cartesian plane.</td>
<td>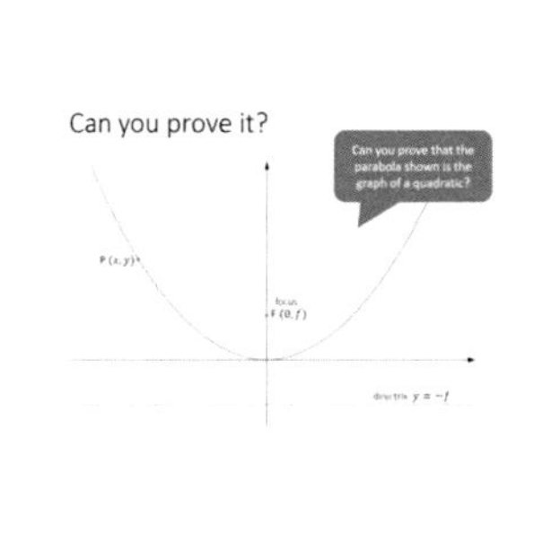
</td>
</tr>
</table>

Support and extension

The templates are numbered in order of difficulty. Encourage more confident learners to create their own designs, either on some of the ready-made templates, or by constructing their own starting grids on squared paper. Some learners may be interested in researching the work of geometric artist Andy Gilmore who draws parabolic designs in this way.

Further ideas

There is plenty of scope in this activity for a cross-curricular project with the Art or Design Technology departments. Learners can create string art using hardboard, nails and string. Or they can explore curve stitching with needles and thread. The axes templates can be printed onto card to provide a more robust surface for this activity.

Learners can also be set the challenge of constructing a giant parabolic design on your classroom display boards using drawing pins and coloured string.

References

Department for Education (2013). *National curriculum in England: mathematics programmes of study*. Crown Copyright.

Acknowledgements

I am grateful to Ken Wessen (@Mr_Wessen) for permission to include the link to his Hungry Bugs app at www.thewessens.net; to Paula Beardell (@PaulaKrieg) for allowing me to share her Phi-scaled hexagons project at www.bookzoompa.wordpress.com; to Jonathan Hall (@StudyMaths) for permission to link to his cardioid tool at www.mathsbot.com; and to Kim Pitchford (@Ms_Kmp) for allowing me to use her curve of pursuit images. The SMP Impossible Objects instructions also provided inspiration.
My thanks to Rachel Smith and Rebekah Mellor-Read for their valuable feedback; to Judith Grandi for editorial assistance and encouragement; and to my son Joe for his sound advice.

Image Credits

Page 3:	© elfinadesign/Shutterstock.com
Page 3:	© upabove/Shutterstock.com
Page 6:	© Utente Stesso [Public domain]
Page 6:	© Daniel Walsh www.danielwalsh.tumblr.com
Page 6:	© Aunti Juli CC BY-SA 2.0
Page 9:	© Tobias R [Public domain]
Page 9:	© Fedorov Oleksiy/Shutterstock.com
Page 9:	© Iryna Kuznetsova/Shutterstock.com
Page 9:	© XYZ/Shutterstock.com
Page 12:	© Юкатан CC BY-SA 4.0
Page 12:	© Dicklyon [Public domain]
Page 12:	© Isroi/Shutterstock.com
Page 16:	© Roman Sotola/Shutterstock.com
Page 16:	© AnonMoos [Public domain]
Page 16:	© MatiasDelCarmine/Shutterstock.com
Page 16:	© Internet Archive Book Images
Page 16:	© Mohammad Fahmi Abu Bakar/Shutterstock.com
Page 16:	© Wellcome Images CC BY-SA 4.0
Page 19:	© Wojciech Swiderski CC BY-SA 3.0
Page 19:	© Zorgit CC BY-SA 3.0
Page 22:	© Firmatography CC BY-SA 2.0